INSTRUCTIONS

FOR THE

CARBINE AND PISTOL

EXERCISES

FOR

THE CAVALRY.

ADJUTANT GENERAL's OFFICE, HORSE-GUARDS,

24th December, 1819.

By Authority:

The Naval & Military Press Ltd

Published by the
The Naval & Military Press
in association with the Royal Armouries

Unit 10 Ridgewood Industrial Park,
Uckfield, East Sussex, TN22 5QE
Tel: +44 (0) 1825 749494
Fax: +44 (0) 1825 765701

MILITARY HISTORY AT YOUR FINGERTIPS
www.naval-military-press.com

ONLINE GENEALOGY RESEARCH
www.military-genealogy.com

ONLINE MILITARY CARTOGRAPHY
www.militarymaproom.com

ROYAL
ARMOURIES

The Library & Archives Department at the Royal Armouries Museum, Leeds, specialises in the history and development of armour and weapons from earliest times to the present day. Material relating to the development of artillery and modern fortifications is held at the Royal Armouries Museum, Fort Nelson.

For further information contact:
Royal Armouries Museum, Library, Armouries Drive,
Leeds, West Yorkshire LS10 1LT
Royal Armouries, Library, Fort Nelson, Down End Road, Fareham PO17 6AN

Or visit the Museum's website at
www.armouries.org.uk

GENERAL ORDER.

HORSE-GUARDS,
24th December, 1819.

His Royal Highness the Commander in Chief has been pleased to direct, that the following Instructions and Regulations for the Exercises of the *Carbine* and *Pistol* shall be practised by the several Regiments of Cavalry.

It is directed in the Regulations for the Formations and Movements of the Cavalry, that " *Regiments, while dismounted, and acting on Foot, shall follow, in every respect, the Rules prescribed for the Battalion in the Infantry Regulations.*"

As soon, therefore, as the Cavalry Recruit shall have been sufficiently instructed in the elementary Exercises of Marching, Facing, &c., he is to be taught the Exercise of the Carbine *on Foot*, and to be carefully instructed in all the details connected with Priming, Loading, and Firing with Ball, the whole of which are equally essential to be understood by the Dragoon, as by the Infantry Soldier.

The Exercises of the Carbine and Pistol *on Horseback* are to be commenced, when the Recruit shall have made a sufficient progress in Horsemanship, as directed in the *Instructions on Military Equitation.*

By Command of
His Royal Highness
The Commander in Chief,

HARRY CALVERT,
Adjutant-General.

CONTENTS.

PART I.

CARBINE EXERCISE ON FOOT.

PART II.

CARBINE EXERCISE ON HORSEBACK.

PART III.

THE PISTOL EXERCISE

Part First.

CARBINE EXERCISE

ON FOOT.

Part First.

CARBINE EXERCISE

ON FOOT.

MANUAL EXERCISE.

THE Recruit, having his carbine slung by the swivel, is to take it in his right hand near the lock, his little finger touching the feather-spring, holding it by his right side, at the full extent of the arm. This position, whether the carine be slung, or not, is called, " *the Trail.*"

Words of Command.	EXPLANATION.

ADVANCE ARMS.

1st Motion.—Raise the carbine with the right hand as high as the hip, and seize it with the left at the *Gripe*, (that is with the full hand round the barrel and stock, with the little finger touching the feather-spring of the lock); the lock downwards; the muzzle a little raised, but

Words of Command.	EXPLANATION.

straight to the front; then *unspring*, by dis-engaging the carbine from the swivel, and seize the small of the butt with the right hand.

2d Motion.—Drop the carbine to the position of the " *Advance*," steadying it with the fingers of the left hand ; the arm square across the body. At the " *Advance*," the carbine is supported by the right hand against the right side, the three last fingers under the cock, the fore-finger under the guard, and the thumb above the guard.

3d Motion.—Bring the left hand to its place on the left side.

SECURE ARMS.

1st Motion.—The carbine is raised about two inches by the right hand, and the muzzle is brought forward from the " *Advance*" about four inches ; at the same instant the left hand is brought across the body, and seizes the car-bine at the gripe.

2d Motion.—Carry the carbine across the body to the left side, and bring it down under the left arm to the position of " *Secure ;*" at the same time withdraw the right hand.

ADVANCE ARMS.

1st Motion.—Lower the butt, and raise the muzzle of the

Words of Command.	EXPLANATION.

carbine to a perpendicular position against the left side; the right hand, across the body, lays hold of the small of the butt.

2d Motion.—Carry the carbine down to the "*Advance*," and steady it with the fingers of the left hand.

3d Motion.—Bring the left hand to its place on the side.

PRESENT ARMS.

1st Motion.—Raise the carbine from the "*Advance*," and seize it with the left hand, as directed for the first motion of the "*Secure*."

2d Motion.—Bring the carbine up to the "*Poize:*" this is done by the right hand elevating it, so that the left being placed along the stock, with the wrist against the guard, and the fingers and thumb pointing upwards, the thumb will be opposite to the left eye. In this position the elbows should be closed to the body and to the butt of the carbine : the barrel is to be upright.

3d Motion.—The carbine is to be brought down perpendicularly to the "*Present*," by the left hand seizing it at the gripe as it descends, and supporting its weight; the butt to cover the left knee, and sunk so low down that the forefinger and thumb of the right hand can lightly grasp the small of the butt, below the guard; the remaining three fingers stretched obliquely downwards, without constraint to the arm or altering the position of the body: As the carbine is lowered, the right foot is drawn back six inches to the rear, but the knees remain straight.

ADVANCE ARMS.

1st *Motion.*—Carry the carbine to the position of " *Advance*," steadying it with the fingers of the left hand.

2d *Motion.*—Bring the left hand to its place by the side.

PORT ARMS.

At one motion throw the carbine to a diagonal position across the body, the lock to be outwards, and at the height of the breast, the barrel opposite the left shoulder. The right hand grasps the small of the butt, just below the right breast; the left holds the carbine at the gripe, the thumbs of both hands pointing towards the muzzle.

N. B.—In this position the pan is opened or shut at one motion, for the purpose of inspection of the locks, flints, &c.

ADVANCE ARMS.

1st *Motion.* Bring the carbine down from the " *Port* " to the " *Advance* " at once; the left hand steadying it.

2*d Motion.*—Bring the left hand to its place by the side.

SUPPORT ARMS.

The right hand is to be brought forward and raised, retaining its hold of the carbine as at the " *Advance,*" until the thumb touches the buttons of the jacket ; the arm to be kept near the body, and the guard of the carbine turned a little to the front.

STAND AT EASE.

In standing at ease, with the arms at the "*Support,*" the left hand may be brought across the body, and laid over the right. It is to be quickly brought down at the word " *Attention.*"

CARRY ARMS.

Drop the carbine at once to the " *Advance.*"

SPRING CARBINE.

1*st Motion.*—The carbine is raised from the " *Advance,*" by the right hand as high as the hip, with lock turned downwards, and is seized by the left at the gripe.

2*d Motion.*—The carbine is *sprung* by the right hand seizing the swivel, and securing it through the ring.

| *Words of Command.* | EXPLANATION. |

3d Motion.—The right hand seizes the carbine at the gripe, and drops it down to the " *Trail,*" at the full extent of the right arm, the muzzle elevated, but directed straight to the front; the left hand quits it at the same time.

PLATOON EXERCISE.

———•———

THE squad, being at the " *Advance*," will receive the commands and instruction for the Platoon Exercise.

Words of Command.   EXPLANATION.

MAKE READY, *(Pronounce " Ready.")*

> By a brisk motion, throw the carbine at once to the position of " *Recover*," in which the left hand holds it at the gripe, and the right at the small of the butt ; the barrel upright, and opposite to the right cheek: then quickly place the thumb of the right hand upon the cock, the fingers under the guard, with the elbow raised, and cock the carbine, dropping the elbow at the same instant.

PRESENT, *(Pronounce " P'sent.")*

> Bring the carbine down to the " *Present*," and look steadfastly along the barrel, at the same time stepping back with the right foot six inches to the rear; place the fore finger before the trigger, but avoid touching it.

FIRE.—By the action of the finger *only*, and by a gradual but firm pressure, pull the trigger, and remain looking along the piece, and wait for the next command.

LOAD.

1*st Motion.*—Drop the carbine down to the *priming position,* and at the same time bring up the right foot to the left heel. In the *priming position,* the carbine is to rest against the hollow of the right side, the muzzle raised as high as the peak of the helmet or chacos, but pointing directly to the front; then seize the lock with the fore finger and thumb of the right hand.

2*d Motion.*—Half-cock the carbine, keeping hold of the cock.

HANDLE CARTRIDGE.

1*st Motion.*—Carry the hand to the pouch, and take hold of a cartridge.

2*d Motion.*—Draw it out, and bite off the end of it.

PRIME.

1*st Motion.*—Shake some powder into the middle of the pan, but not more than will *half* fill it; place the last three fingers behind the steel, holding the cartridge between the thumb and fore finger.

2*d Motion.*—Shut the pan, and seize the small of the butt between the last three fingers and the hand.

LOAD.

1st Motion.—Turn the carbine smoothly round to the *loading position*, in which the barrel is turned towards the front, the toe of the butt rests against the outside of the left leg; the muzzle pointed forward, and opposite to the middle of the body: the right hand holding the cartridge, is placed against the muzzle, covering the sight.

2d Motion.—Shake all the remaining powder into the barrel, then put in the paper as wadding, or the ball, take hold of the head of the ramrod with the fore finger and thumb.

DRAW RAMROD.

Draw out the ramrod, and put it an inch into the barrel, with the arm extended.

RAM DOWN CHARGE.

1st Motion.—Push the cartridge to the bottom.

2d Motion.—Strike it twice smartly with the ramrod.

RETURN RAMROD.

Draw the ramrod out of the barrel, and return it into the pipe without loss of time, forcing it well home; then face to the proper front, the fore finger and thumb still holding the head of the ramrod.

Words of Command.	EXPLANATION.

ADVANCE ARMS.

Throw the carbine across the body, at one motion, to the " *Advance,*" and instantly quit the left hand.

Explanation of the Command, " *Prime and Load.*"

PRIME AND LOAD.

1st Motion.—Bring the carbine, by one brisk motion, to the priming position, the thumb before the steel, the fingers closed, and the elbow a little turned out, in order that the wrist may be clear of the cock.

2d Motion.—Open the pan by a close motion of the right arm, dropping the elbow down, and extending the hand along the front of the lock.

3d Motion.—Carry the hand to the pouch and seize the cartridge.

The rest of the loading Motions as above described, except that they are to be made with as much dispatch as possible, every man returning his piece to the " *Advance,*" or " *Recover,*" immediately.

Explanation of the Position for each rank in Firing.

Front Rank.

Cavalry, when dismounted, is always to be formed in two ranks. The front rank man, having his carbine at the *Recover*, and *Cocked*, is to bring it down to the *Present*, stepping back six inches to the rear with the right foot. After having fired, the right foot is to be brought up square with the left, and the priming and loading proceeded with as before directed.

Words of Command.	EXPLANATION.

READY.—At the " *Recover*," as before directed.

PRESENT.—As before.

FIRE.—As before; and, when fired, remain looking along the barrel at the object aimed at, until " one, two," may be distinctly told; then proceed to prime and load without loss of time.

Rear Rank.

READY.—The *Rear Rank*, on receiving the command " *Ready*," cocks at the *Recover*, taking a moderate pace to the right, with the *right foot only*.

PRESENT.—As before.

FIRE.—As before directed.—On bringing down the carbine to the priming position, close the left foot to the right, preserving the quarter face to the right; then proceed, as before directed, with the priming and loading Motions. When these are completed, take ground to the left, and cover the front file.

END OF PART FIRST.

Part Second.

CARBINE EXERCISE

ON HORSEBACK.

Part Second.

OF THE USE OF THE CARBINE AND PISTOL ON HORSEBACK.

WHEN the Recruit has attained a degree of proficiency on foot, the Exercise of Arms on Horseback should, in general, form a part of each riding lesson; by this means he will acquire such dexterity in the use of his Fire-arms, as will enable him to load and to discharge them, while his horse is in motion, without annoying the animal, or being disturbed in his seat.

Although it is desirable that the horseman should be habituated to the use of his Carbine at speed, few occasions can arise for his using it against an enemy at any pace beyond a walk; and notwithstanding he is enjoined, while skirmishing, to keep his horse in motion, in order to avoid becoming a fixed object for the enemy's marksmen, he cannot reasonably calculate on his shot being effective, unless he halts for the moment of firing.

It is found that the fire of the *Carbine* to the *left*, and of the *Pistol* to the *rear*, are the most effective; and that to the *right* with the *Carbine*, and to the *front* with the *Pistol*, is the least so. It should, however, be remembered, that although the fire of the carbine be most certain to the left, the turning of either flank to the enemy, exposes both man and horse in the greatest degree.

The fire of the cavalry soldier is never to be had recourse to but in skirmishing ; and, as in that situation, it may be considered rather as a demonstration than an effectual mode of offence, firing with the carbine to the *front* is generally to be preferred ; because in that position the horse presents the least mark, and the rider is most covered from the shot of the enemy.

The Pistol is ill calculated for skirmishing against the enemy : when on such occasions the carbine cannot be used, recourse must be had to the sword.

There are, however, situations of emergency where the horseman may find the pistol useful ; as when his sword is broken, or his sword-arm partially disabled : If under these, or similar circumstances, he should be compelled to make a precipitate retreat, he may, by presenting his pistol, keep his enemy at bay ; although it would seldom be advisable for him to fire until his adversary should close upon him, and the effect of his fire would be morally certain.

The following instructions are given in detail as they are to be taught to Recruits in small squads ; but as soon as they are perfected in the performance of the several motions by command or signal, they may be taught to proceed with the execution of the several commands, without loss of time ; and afterwards the Recruit may fire blank cartridge.

In the first essays of the Recruit in this part of the exercise, great care must be taken, that in presenting to the front or left, he does not strike or touch the horse's head with the carbine or pistol ; and in firing, that his horse's ears be not singed or struck by any loose grains of powder from the pan. These accidents may be avoided by raising the breech of the piece sufficiently high, and by turning the lock a little upwards at the Present.

In *Priming*, the Recruit must be made to understand that a small quantity of powder in the middle of the pan is sufficient; that he must never fill it, or scatter, or leave any loose grains on the edge of it, as by that means the hammer would be prevented shutting down close, and the priming would be lost.

In *Loading*, he must be taught to shake all the remaining powder out of the paper into the barrel before he puts in the wadding; and when loading with ball, to double the paper round it, so that it may require a small degree of force to drive it home to the charge, otherwise when he may have occasion to " *sling* " or " *strap* " his carbine, or " *return* " his pistol, when loaded, the ball would be apt to fall out.

When the Recruit is familiar with the firings at the halt, he may practise them while his horse is in motion; afterwards he must be taught to fire with ball at a suitable object, first at the halt, and afterwards, when in motion.

In all the motions connected with firing, great care must be taken to avoid altering the accustomed feeling of the bridle in the horse's mouth, or the usual seat and balance of the man, as tending to alarm the animal; for a horse once rendered timid by an accident in firing from his back, will make the practice of it both difficult and dangerous.

THE

CARBINE EXERCISE

ON HORSEBACK.

THE Squad for instruction is to be formed in a rank-entire at double open file distance, and a flugelman in front to give time.

Words of Command.   EXPLANATION.

SPRING CARBINES.

1st Motion.—Take off the right hand glove and the lock cover, and put them into the shoe-case; seize the carbine at the small of the butt, with the back of the hand upwards.

2d Motion.—Bring the butt forward, until the carbine be nearly upright; take hold of it with the fore-finger and thumb of the left hand (still holding the reins), and with the right seize the swivel, placing the thumb on the spring.

3d Motion.—First spring, and then unstrap the carbine; quit the hold of it with the left hand, and seize it with the right at the gripe.

4th Motion.—Draw the carbine from the bucket, and continuing to grasp it in the full hand, let it rest in the hollow of the thigh; the barrel diagonally across the body; the muzzle a little elevated, so that it be in line with the horse's left ear.

This position is called the " *Advance.*"

In this position the carbine is carried by small detachments and advanced parties when near the enemy, and by videttes on service, being that from which the dragoon most readily prepares to fire, and which occasions the least fatigue.

PRIME AND LOAD.

Place the carbine in the full of the left hand, at the gripe (without disturbing the position of the arm, or the feeling of the bridle in the horse's mouth), keeping the carbine in the same diagonal direction as the *Advance;* place the thumb of the right hand before the steel, or hammer, the elbow a little raised.

This is called the *Priming* position.

PRIME.

1st Motion.—Open the pan, carry the hand round to the pouch, and take hold of a cartridge.

2d Motion.—Draw out the cartridge, and bite off the end of it.

3d Motion.—Shake a little powder into the pan, and with the three last fingers shut it, then seize the small of the butt.

LOAD.

1st Motion.—Raise the carbine with both hands (without altering the position of either upon it), clear over the hollow between the holsters and the horse's neck, and carry the butt under the bridle-reins, to the near side (called *Casting About*), letting the carbine turn in both hands till the lock be to the left; then permitting it to slide through the left hand until the muzzle be opposite to the right breast, the right hand is brought up to the sight. In this position the carbine will be sustained principally by the swivel.

2d Motion.—Shake the powder into the barrel, and then put in the paper, or ball, and lay hold of the ramrod with the fore-finger and thumb.

RAM DOWN CHARGE.

1st Motion.—Draw out the ramrod, and put an inch of it into the muzzle.

2d Motion.—Ram down the charge, and drive it home by two distinct beats of the ramrod.

3d Motion.—Return the ramrod and hold it between the fore-finger and thumb.

ADVANCE CARBINE.

Raise the carbine up with the left hand, and seize it at the gripe with the right, carry it over the horse's neck, and place it at the " *Advance;*" the bridle hand resumes its position.

MAKE READY.   (Pronounced " *Ready.*")

Place the carbine in the left hand in the *priming position;* cock; then seize the small of the butt with the right hand.

TO THE FRONT PRESENT.

Raise the carbine to the *Present,* with both hands, and place the butt firmly against the hollow of the right shoulder ; lean the head in order to take a steady aim.   In raising the carbine to the *Present,* care must be taken not to disturb the feeling of the bridle in the horse's mouth ; and with this view, the motions must be made as smoothly and quietly as possible : the body may lean a little forward, and, if necessary, the reins may be a little lengthened.

FIRE.

Pull the trigger, still keeping the carbine at the *Present,* and the eye fixed on the object.

PRIME AND LOAD. ⎫
RAM DOWN CARTRIDGE. ⎬ As before explained.
ADVANCE CARBINE. ⎭

N. B.—Preparatory to firing to the left, the men must be made to turn their horses to the right, in order to avoid injuring each other.

MAKE READY.—As before explained.

TO THE ⎫ Raise the carbine to the *Present,* to
LEFT ⎬ the left, with the right hand ; and in order
PRESENT. ⎭ to steady it and ensure a good aim, rest the
barrel on the left arm, just above the elbow, which for this purpose is to be raised nearly as high as the shoulder.

FIRE. ⎫
PRIME AND LOAD. ⎬ As before explained.

N. B.—Preparatory to firing to the right, the men must turn their horses to the Left About.

MAKE READY.—As before explained.

TO THE ⎫ Turn the body to the right, but without
RIGHT ⎬ deranging its balance, and raise the carbine
PRESENT. ⎭ to the *Present,* with the right hand, placing
the butt firmly against the hollow of the shoulder. The bridle-hand is to preserve its usual position.

Words of Command.   EXPLANATION.

FIRE.
PRIME AND LOAD. } As before explained.

From the *Present* it may be occasionally necessary to suspend the firing, by command or otherwise.

RECOVER ARMS.

    Carefully withdraw the finger from the trigger, and let the carbine drop gently down to the *priming* position in the left hand.

From the Recover, the carbine may be again *Presented,* or may be *Half Cocked, Advanced,* &c.

From the *Advance* the carbine may be *Carried* or *Slung.*

CARRY CARBINES.

    Without altering the position and grasp of the right hand, raise the carbine and place the butt of it in the hollow of the thigh, where the hand previously rested ; the muzzle to be carried to the right, so as to be in a line, clear of the horse's neck, on that side, and leaning, rather forward, the elbow near the side.

In this position the carbine is carried by the advanced men in marches of parade, through towns, or the relief of guards on home service, &c.; but it ought to be carefully avoided when near an enemy, as it has often happened that the post of a vidette, the march of patroles, &c., have been discovered by the glistening of the light on the barrels of the arms, which would probably not have been seen if they had been carried in a less conspicuous position.

SLING CARBINES.

This position is taken from any of the preceding, by quietly dropping the carbine with the muzzle downwards behind the thigh, and leaving it *Slung,* or suspended by the swivel only.

From being *Slung,* the carbine may be brought at once to any of the foregoing positions, or may be *Strapped.*

STRAP CARBINE.

Seize the carbine at the gripe, and fix the muzzle in the bucket; then take hold of it with the fore-finger and thumb of the left hand, as in *Springing;* first strap and then unspring the carbine, and dropping the swivel, both hands resume their usual positions.

END OF PART SECOND.

THE

PISTOL EXERCISE.

Part Third.

THE

PISTOL EXERCISE.

THE squad, being mounted, is to be formed as for the Carbine Exercise.

Words of Command.　　　　　EXPLANATION.

DRAW PISTOL.

1st Motion.—Take off the right hand glove, unbutton the flounce, and push forward the cloak, or draw back the sheepskin and shabraque, according to the equipment, and seize the butt of the pistol with the right hand under the left arm.

2d Motion.—Draw the pistol carefully, and bring it at once to the position in which the sword is *Carried*, the muzzle upright, the cock resting in the hollow between the thumb and the hand, the lower fingers relaxed and extended along the butt. This position is called the *Advance*.

Words of Command.   EXPLANATION.

PRIME AND LOAD. ⎫ These several motions are to
RAM DOWN CHARGE. ⎬ be made in the same manner
MAKE READY. ⎭ as directed for the carbine.

TO THE ⎫ From the left hand raise the pistol with the
FRONT ⎬ right, till the breech be nearly as high, and
PRESENT. ⎭ in line with, the right eye, with the muzzle
lowered to the object; the hand lightly
grasping the butt, the arm a little bent, and
without stiffness, in order to keep the
pistol more correctly to its aim, and to avoid
the shock of a recoil.

FIRE.
PRIME AND LOAD. ⎬ As before directed.

Preparatory to firing to the right, or left, the squad must
turn their horses, as directed in firing to a flank with the
carbine.

MAKE READY.—As before.

TO THE LEFT, ⎫ Resting the muzzle on the left arm, as
PRESENT. ⎭ before directed for the carbine.

PRIME AND LOAD.—As before.

TO THE RIGHT, ⎫ The pistol is carried to the right, is
PRESENT. ⎭ raised and levelled as directed in
presenting to the front.

FIRE.
PRIME AND LOAD. } As before.

TO THE REAR, } Carry the pistol as far towards the rear
   PRESENT.   as the body, turned in that direction,
will admit; take the aim, and hold
the pistol in the same manner as directed for presenting to the front.

FIRE, *&c.*—As before.

RETURN PISTOL.

1st Motion.—Drop the muzzle under the bridle-arm, and
place the pistol carefully in the holster.

2d Motion.—Bring the right hand to its position by the
thigh.

END OF PART THIRD.

www.ingramcontent.com/pod-product-compliance
Lightning Source LLC
Chambersburg PA
CBHW020953030426
42339CB00004B/80